Some Alternative Pathways

for the

Hesitant Queen Rearer

Northern Bee Books

Some Alternative Pathways for the Hesitant Queen Rearer

© Ben Harden

ISBN 978-1-908904-55-3

Published by Northern Bee Books, 2014
Scout Bottom Farm
Mytholmroyd
Hebden Bridge HX7 5JS (UK)

Design and artwork: D&P Design and Print

Printed by Lightning Source, UK

Front cover: Spring gentian, *Gentiana verna*, growing in The Burren, Co. Clare.(It's common habitat is in high alpine meadows rather than near sea level as in The Burren)

Some Alternative Pathways

for the

Hesitant Queen Rearer

Northern Bee Books

CONTENTS

1. INTRODUCTION

There are many fine books on queen rearing which assume quite a degree of skill, not wishing to emulate them the aim of this addition to the subject is to show ways of overcoming aspects of the processes that some find daunting. Where a large scale operator needs speed from skill and experience the smaller scale hobbyist can afford to spend time and be patient.

There is a magic in queen rearing where human manipulations change what was destined to be a humble worker bee into a fully fledged queen. It must be remembered however that mere humans are not as good as fairy godmothers at magic so things do not always work out as hoped for; or as they did before. This applies to professional queen producers just as it does to amateurs. When we encounter failures the tendency is to blame ourselves whereas it can be the bees who say NO. An example can be seen when a colony that had initiated swarm preparations changes its mind and tears down the queen cells of its own accord.

The skills normally deemed necessary of a queen rearer are an ability to find queens and also to be able to see eggs. There are ways and means around both. Starting with seeing eggs, for those who have difficulty the first step is to try the magnifying aids sold for DIY purposes. These can be worn under the veil can be flicked up or down and some even have led lighting of the subject being examined. For those who have difficulty finding queens the necessary steps to bypass the need will be apparent later on. As practice improves ones skills and fosters confidence, so it also leads to greater ability in locating a queen. A perceived inability to find a queen is not necessarily an excuse for putting off queen rearing.

It has long been known that when bees find themselves queenless they will immediately start to raise a replacement from a worker larva and this is the basic mechanism that is utilised when queen rearing. Over time many methods have evolved from tweaking what went before and on this basis it has been found that queen right colonies can also be induced to raise queens without swarming.

The sequence in the formation of a queen honey bee is for the larva of a female egg which is in a cell hanging vertically to be lavishly fed with a richer diet than offered to a worker and so grow into a queen larva rather than a worker larva. When bees raise a queen themselves a female egg is laid in a queen cup which hangs vertically and hatches after three days. An egg in a worker cell also hatches after three days. These freshly hatched larvae are initially fed an identical diet of royal jelly and so are identical, worker and queen. The only difference being the positioning of the cell, vertical for a future queen and sloped gently upwards for a worker. The diet of these larvae alters after the initial feed with the queen larvae being fed more lavishly and on a higher plane of nutrition than the worker larvae. If a freshly hatched larva in a worker cell is placed in a vertical cell in the right position in a colony and fed as a queen larva it will develop as a queen. This just hatched worker larva being identical to a queen larva at that stage when fed as a queen becomes a perfect queen. As a worker larva ages and its diet changes relative to a queen larva of the same age its chances of becoming a perfect queen decline.

All of this in practice means that when rearing queens artificially the beekeeper should start with freshly hatched worker larva to ensure the best results. The aim should be for larvae that are up to 12 hours old. If a little older it is not a total disaster, the queen raised might not be the absolute best but likely to be still very acceptable. It is all part of mastering one of the facets of queen rearing. If old larvae from two days plus are used then results get unpredictable some carry on as workers, some become queens and some inter casts that is an individual betwixt and between worker queen. Bees left to themselves can identify the older worker larvae that will still make full queens and when forced to, will raise a queen on a worker larva up to three days old, so beware if breaking queen cells in attempts to control swarming, two days after inspection the bees can have a sealed queen cell and go.

As a basis to work on the development of a queen bee can be placed in the following stages,

1.   A female egg is laid which hatches into a larva after three days and is fed a queen's diet of Royal Jelly

2. The larva grows and it's cell is sealed after a further five days. Within the sealed cell the larva grows and metamorphoses into an adult queen bee

3. Sixteen days after the egg was laid the queen hatches and the queen takes her mating flights

4. The queen goes on to head a colony

These stages can now be examined one by one.

2. THE FEMALE EGG AND LARVA SELECTED FOR A FUTURE QUEEN.

When raising queens the first decision is the selection of a breeder queen and here colony records should be the basis of that decision. It is then necessary to decide on timings, bees do not raise queens out of season so we are confined to the time from early summer through to early autumn and the local experience of others will, if necessary, put dates on that time span. The whole process is governed by a biological clock so start date dictates when the later stages are reached and all will fall asunder if some family/social commitment prevents a beekeeper doing the necessary on the specific day. The dates for each stage need to be put in the diary and kept.

There are commonly two methods of getting a freshly hatched larva from the breeder queen.

Queen Rearing kits

Firstly the queen can be confined in a queen rearing kit where she is forced to lay in artificial worker (female) size cells which when removed become artificial queen cell cups. There are two such kits commonly used one made by "Jenter" and this provides the common use name for such kits and a second made by "Nicot" commonly called "Cupkit" or "Cupularva". In both the kit is secured in a full brood frame and set up as per maker's instructions. The important thing with these is to get them smelling right for a bee and this should be done before use each season, this operation is referred as "familiarising" and is achieved by placing the frame in an active brood chamber for a period of some days. It is of benefit to douse the kit with syrup for the bees to clean off and in the case of the Jenter the bees also need to draw comb on the plastic foundation of its base before queen introduction. The chosen breeder queen is found and shut into the kit, she has no option but to lay in the cells provided and is released back into the brood chamber some 12 hours later. Provided the queen is in full lay there should be no hitches as she can hardly stop egg production to the point that queens are sometimes found when the egg protruding from the abdomen is what first catches the eye. Three days plus a few hours later the frame carrying the kit is removed and all going well will contain many freshly hatched larvae. These are seen at the base of the cell lying on a glistening bed of royal jelly as in the photo. There are usually many more larvae than are required so remove as many as needed and place onto cell bar/s. It is then a management decision as to what is done with that frame with its queen kit and the larvae and stores it holds.

The advantages of kits are that the age of these larvae is known and they are readily transferred on to a cell bar for the next phase of rearing. The disadvantages are that the kits are expensive and require the queen to be found, shut in and then released putting an extra entry in the diary when things must be done.

Grafting

The next method is grafting and here there are perceived ideas that 20/20 vision, skill and dexterity is necessary which is not necessarily so. Starting with 20/20 vision the magnifying aids sold for fine DIY use will help enormously. Once over the fear, there is not much skill needed and if dexterity is the big barrier then look laterally. Pictures of large scale queen rearers, whose livelihoods depend on getting it right, be it for beekeeping or royal jelly production show an individual grafting while sitting comfortably at a table in a room with no bees about. This person need not be a beekeeper at all so an outside helper or family member can do the grafting for you. Even to the point of somebody who has no wish to go anywhere near a bee hive be it open or closed.

When looking for recently hatched larvae to graft, go through the breeder queen's brood chamber looking for eggs and then look to see if rather than a full expanse of eggs there are those interspersed cells with glistening material in them. These will be young larvae on their bed of royal jelly, think of a row of peas sown in the garden, they do not all emerge simultaneously and in similar manner eggs do not hatch sequentially so here is where the fresh hatch is to be found. Brush the bees off this frame rather than shaking them so as to leave the larvae central in the cell bottom rather than shaken into the bottom corner of the cell where it is more difficult to access them. Now wrap this frame in a damp towel and take it to the grafting venue.

An immediate question is. Won't this period outside the hive environment of temperature and humidity harm the larvae? The answer is no provided extremes are avoided. Bees and their brood are poikilothermic meaning their temperature is taken from the environment they inhabit but within limits. Open brood does not chill in working day time temperatures but it does bake when exposed to strong sun so be wary having frames exposed to strong sun for any length of time. There is also a danger of killing brood by allowing it to dry out, however this is not a major concern in temperate climates. When we are given the development times for the various stages in a bee's progress from egg to adult these are averages, the cooler it is the longer the development and the warmer the shorter so in essence development time is a factor of so many degree hours. By holding open brood cool and moist its metabolic rate is slowed down resulting in an elongation of just that period of development.

I often bring a frame back from an out apiary wrapped in a damp towel for grafting at home. On one occasion such a frame then went on to the association apiary for return the following day when it was placed in a hive and the brood progressed normally from then on. At Rothamstead on one occasion a brood frame was inadvertently left over night propped against a tree. The next morning grafts were taken from it with no ill effects on take or subsequent queens. (Norman Carreck, personal communication) These two examples are extreme but show that some hours out of a hive environment do no harm to brood when it is neither over heated nor dried out.

The technique of grafting is, in the ideal demonstrated. If an apiary meeting can be convened with somebody to show how it's done so much the better and the various grafting tools should be on hand as different individuals have differing preferences. A point to bear in mind is that is easier to pick up a larva than to place it in a queen cup. Other than in dry practice runs the cell bar that carries the grafts should be "familiarised" before taking the grafts and this is done by covering in syrup and placing in the raising colony for some 24hours before the grafting.

The ideal larva to graft is under 12 hours old and found adjacent to and amongst unhatched eggs. It is small and looks like a lower case c. Experience leads to easy recognition of the right size but is not a disaster if in early runs larva slightly older than the ideal are used. These will make reasonable queens, not quite the best, but it all helps in mastering techniques and a realisation that queen rearing is possible. Nobody rode a bicycle with no hands on the handlebars on day one!

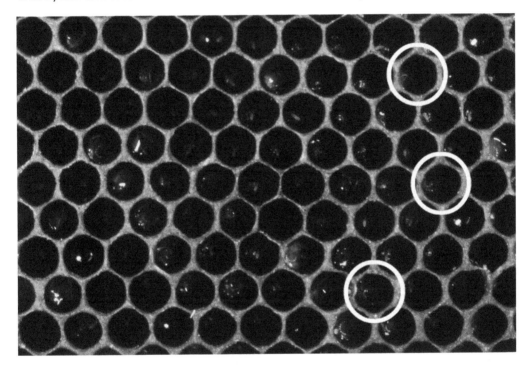

The white circles indicate the target size for larvae to graft. Slight movement of the frame or head would show up more examples than are easily seen here. Some authors incorrectly claim that eggs lean over progressively from standing up on the cell base on day 1 to lying prone on day 3 when they hatch. Eggs do not move until the hatching process starts on day 3. (DuPraw E.J. 1961. A Unique Hatching Process In The Honeybee. Transactions of the American Microscopical Society Vol.LXXX (2):185-191)

Grafting Tools

1) Stainless steel which is expensive but feels marvellous. The leading edge is quite blunt in relation to the size of a larva and benefits

from rubbing down with emery paper. It is one of the more difficult tools to remove the larva from.

2) A water colour paint brush 000 or 0000 size. In use dampen the bristles and slide down the cell wall and under the larva to lift it. To unload place the larva in the bottom of the queen cell and roll the brush away from under it.

3) A "Chinese" grafting tool. Again slide the quill down the side of the cell and under the larva to lift it. To unload press the quill portion on the base of the queen cup and slide the larva off as in the diagram.

4) The classic homemade tool is made from a goose quill or even a quill from a lesser bird. The end is cut back to fashion a narrow strip suitable for sliding under a larva. That late great bee man Steve Taber's choice was a paper clip hammered into his desired shape. There are many possibilities.

Unloading Chinese Grafting Tool

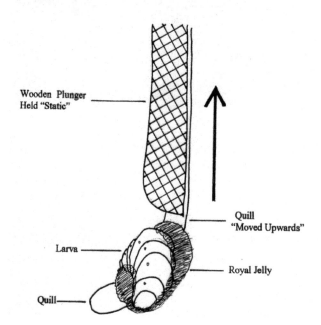

Wooden Plunger Held "Static"

Quill "Moved Upwards"

Larva

Royal Jelly

Quill

The target in grafting is to lift the young larva from a worker cell and place it in a queen cup. It is difficult to see what you are doing with all but the fine stainless steel tool and it is quite legitimate to cut down the worker cell walls with a scalpel or similar craft knife to aid vision. Bees will have no difficulty repairing such damage. With a little practice the need to see precisely how things are going is replaced by knowledge gained from feel so the scalpel can be retired. If when starting out things are not going well and larvae are getting damaged just remember that there are 100's out there so those few are incidental. For some it is a help if a small drop of water is placed in the queen cup first which allows the larva to be floated off the grafting tool. When picking up a larva one school of thought advises approaching from the convex side because when putting it down again the larva stays the same side up. This is easily demonstrated by picking up an opened pair of spectacles from a flat surface with a pencil approaching from each side and then putting them down again. Approach from the convex side, in front, and the spectacles go down the same way up, approach from the concave side, between the arms, and when put down they go upside down.

A Potential Problem! Having charged the queen cups with all but the Chinese tool it is seen that there is a dearth of royal jelly with these larvae. The initial concern is that the larvae will starve but there will still be sufficient until the bees put things to rights again.

Cell Bar Frames

These are ordinary empty brood frames in which one or two strips of wood have been attached to run parallel with the top bar. They are commonly the same width as a side bar 7/8" or 22mm and either readily detachable or held by a central screw each end to allow the bars to rotate. The reason for this is to facilitate attaching the queen cups from a kit or when grafting as the bar can be placed horizontally to work on and then turned vertically for insertion in the hive. The number of queen cups attached to a bar is a matter of beekeeper preference with two bars a brood frame being common. As a rough guide 20 cell cups or so in total is a reasonable maximum for smaller scale operators. Not all will take so allow for wastage. It is possible to buy the plastic cell cups and their fittings separately from the queen rearing kits which is the easiest way to go, there is also the advantage that these fit well later on with mating hives. Besides the cups from the kits there are independent manufacturers of plastic queen cups. With any type of queen cup first off you should check that they are compatible with the mini nuc that will be used later on if that is the chosen path for the queens.

If you enjoy a challenge queen cups can be homemade. Take a short length of 3/8" or 10mm dowel put in the chuck of an electric drill switch on and shape the free end to a nice dome with sand paper. Mark the dowel 3/8" up from this domed end with a pencil line, dip in water and then in molten beeswax up to the line and remove, allow a short time for the wax on the dowel to cool before immersing again in the wax but not quite as far and remove and repeat thus building up wall thickness until satisfied. The wax cup so formed will twist and slide off the dowel. With ingenuity these cups can be stuck with more wax to a bar or lollipop sticks or whatever to aid their use later in the process, all in all a challenge but it can be done.

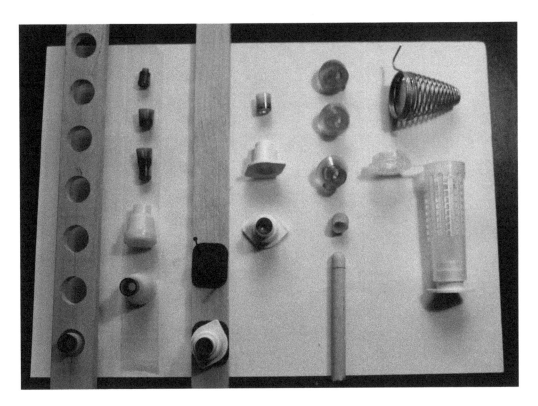

From left to right, colulmns 1&2 Jenter cell bar and cell cups, columns 3&4, Cupkit cell bar and cell cups. column 5 3JZ cell cups, a wax cell cup and a dowel for making wax cups. column 6 a wire cell protector and a "Haircurler Cage"

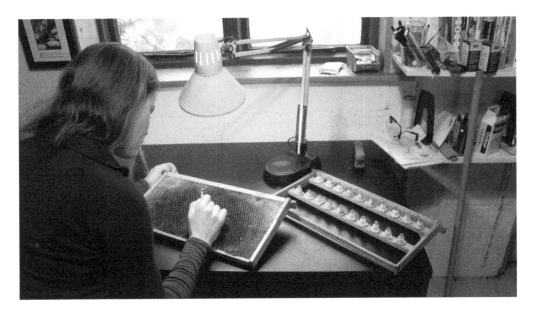

A comfortable bee free grafting bench. As a rule women have a greater facility for fine intricate work than men do so men could consider asking for their wife or daughter's help.

Rather than a perfect or near perfect take of queen cells an example of an acceptance rate that can readily be bettered!

3. QUEEN CELL INITIATION

A colony of bees is totally dependent in having a queen so whenever they find themselves queenless they start to raise a fresh queen provided there are young worker larvae capable of being brought onto queens. This has been the mechanism commonly used by queen rearers in the past and the most reliable application of this principle is with a swarm box. This is a specialised box filled with nurse bees taken from multiple colonies and given frames with pollen and stores but no brood except for the cell bars with charged queen cups. It is for starting more queen cells than most beekeepers wish for so not really a topic for this book.

Queenless colonies can be used to start queen cells but this disrupts the colony and needs careful management.

The producers of Royal Jelly need to get huge numbers of queen cells initiated in order to harvest royal jelly in quantity. In France these beekeepers have developed methods of getting queen cells raised in queen right colonies which must be <u>prosperous and strong.</u> This avoids difficulties with queenless bees as well as permitting continuity of jelly and honey production. Of these methods the most appropriate for small scale or hobbyist beekeepers to adopt is one where standard hive bodies are used and kept in a vertical stack. The set up is to use a double brood chamber with the queen confined in the lower box by a queen excluder. On routine examinations the frames of sealed brood in the lower box are moved into the upper box to be replaced by frames from which the brood has hatched and thus ready for the queen to lay up again. When used to produce royal jelly a frame feeder for use if there is no honey flow, two frames of pollen, a frame of young brood and the cell bar frame are placed in that upper box and the box filled out with sealed brood or stores. English language references are R.F. van Toor, Producing Royal Jelly, published by Northern Bee Books and Giles Fert, Beekeepers Quarterly. 63:26-32

The National Bee Unit in York raise queens on this principle, by allowing the queen cells to go on to be capped. Even though there are sealed queen cells in the hive the bees do not swarm. 2002, Wilkinson D. & Brown M. A. 2002. American Bee Journal 142 (4):270-274.

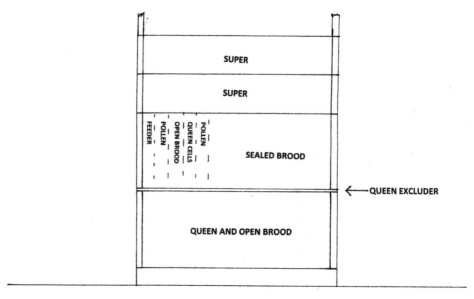

Cell raising when using a double brood chamber colony

For those who keep bees on a single brood chamber there is a modified version where a second brood box is added which only takes the two pollen frames, a frame of young brood and the cell bar frame plus a frame feeder if needed with the rest of the space taken up with large dummies. (Ben Harden, A Simple Method of Raising Queen Cells, Beekeeping in a Nutshell No. 59). An important factor with these three set ups is to bring all, or as much pollen as possible, up close to the frame of cell cups. This is because bees do not move pollen about as they do with liquid stores and the royal jelly secreting bees needing pollen to make royal jelly are thus brought to the frame of queen cells where they are wanted. The frame of young open brood is what first brings the necessary nurse bees up so this should be on one side of the cell cups and the best pollen frame the other.

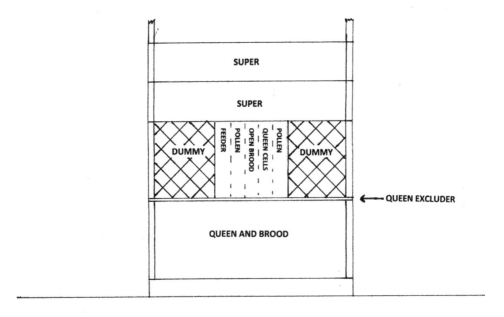

Cell Raising when using a single brood chamber colony

With any of these arrangements the procedure is first to place the cell bar frame in the cell raiser's brood box for a day or so before adding larvae for familiarisation. After this the cell bar frame is removed for filling with larvae in the same operation as the hive is set up as per previous diagrams while remembering the importance of the pollen frames. Once the larvae have been placed in the cell cups the cell bar frame is taken and put in its allocated position in the hive. All going well the bees will accept a proportion of the larvae, the numbers depending on many factors, and build queen cells. Queen cells can be raised in such an arrangement with just a queen excluder separating the brood boxes however some beekeepers like to have greater separation for a day which can enhance the acceptance rate. This seperation can be,

1) A Cloake Board, which is a modified queen excluder having a built in sliding panel to seal off the excluder for the desired period.

2) A full covering of the queen excluder with a sheet of hardboard or plastic which is removed after a day.

3) A compromise where a sheet of plastic is placed in the middle of the excluder leaving a margin of uncovered excluder of 40mm or 1 ½"

all round. For National hives this sheet would be 13 ½" square and such a sheet can be left in situ until convenient to remove it. Once these queen cells are sealed 5 days later they are fully provisioned and continue their development.

Potential difficulties.

Very often these sealed queen cells can be left as they are to mature however there are dangers.

1) A queen cell can be drawn on the frame of young brood brought up to attract nurse bee to the queen cups. If on an older larva that queen will hatch first and kill her rivals so check for and destroy any such cell.

2) The bees can decide to tear the queen cells down.

3) If there is a good flow the bees can encase the cells in comb and honey, the cell cap gets reinforced to the point where the young queen is unable to emerge and so is lost.

To safeguard against such difficulties there are two alternatives. The first, simplest and cheapest is to use a proprietary cell protector known as a "haircurler cage". These are perforated covers which encase the queen cell <u>once it has been sealed</u> and thus prevent the bees access to the cell. If it so happens that a queen hatches earlier than expected she can't get out to destroy the other queen cells but can be fed through the perforations by the worker bees.

The second option is to remove the queen cells once sealed and place them in an incubator. This is a standard incubator for fowl eggs which has been adapted to hold queen cells vertically, i.e. some sort of frame that the queen cells hang from. The temperature is set for 34.5 to 35 degrees Centigrade and the water trough kept filled to maintain humidity levels. Here again cells are often placed in cell protectors in case a queen hatches early or if it has been decided to introduce a hatched virgin into a mating nuc rather than a ripe queen cell. The down side is cost and really incubators are more for the large scale queen rearers.

4. THE HATCHING OF SEALED QUEEN CELLS

1) Provided the queen cells have been placed in hair curler cages they can be allowed to hatch there for introduction as virgin queens. If with bees the workers will feed the virgins in these cages but if in an incubator then a very small drop of honey should be put on the inside of the lid to tide the queen over until released amongst workers.

2) Working with queen cells, these need to be handled carefully as both the legs and wings of a developing queen can be damaged if they hit off the cell wall. Once the developing queen nears maturity both legs and wings have hardened off and cells can be more readily handled with little risk. On this basis sealed cells on day 14 after the egg was laid, or 11 days after grafting or removal from a queen rearing kit need to be transferred to the next phase of the process. If there are doubts about the age of the larvae chosen initially then move a day earlier. It is most dispiriting when a queen hatches out earlier than expected and destroys the rest of the queen cells.

The simplest use of a sealed queen cell be it a natural one or reared, is to put it in a wire cell protector and then ensuring the open top is covered over with the metal slide use the spike to insert it in the middle of an outer brood frame of the colony to be re queened. It doesn't always work but being protected the virgin queen can hatch safely and all going well is treated as a supercedure queen who will go on to lead the colony.

A queenless nuc can be made up and given a sealed queen cell either as above or by making an indent with the thumb in the side of a brood comb then pressing in the queen cell thereby attaching it to the comb. This nuc should be made up about a day before introducing the cell to aid acceptance as by then the bees are aware of their queenless state. If you have difficulty finding queens a queenless nuc can be made up by taking a fresh brood chamber and into this place frames from the donor colony or colonies that have had <u>all</u> the bees shaken off into the brood chamber that they came from. There should be a frame of sealed brood, open brood in all stages and stores plus mixes to make up to the 4 or 5 frames required. This brood box is now returned to the donor hive but above the queen excluder and the hive reassembled with the voids in the queen's brood chamber having been filled with fresh frames. Missing frames in the upper brood chamber are not of concern due to the short time involved, ensure however the 4 or 5 frames are pushed together. Nurse bees from the brood chamber and even some from the supers will come to cover this brood within a quarter of an hour or so and as the queen must still be in the bottom box that upper brood box can be removed with its frames now going into a nuc box and there is a queenless nuc. It is not a problem if such a nuc lacks established foragers as some of the bees therein soon move into that role to fill the void even if younger than they would have done had they stayed in an established colony. Any fresh nuc when made up needs to be taken away some 2 miles or more to stop the older bees when recognising their foraging area returning to the parent hive. All going well, the queen cell hatches the virgin queen mates and goes on to head this colony.

The queen once hatched still needs to mature then mate before starting to lay. As a rule the larger the colony the longer it takes for her to commence laying, the shortest time for this is in a mini nuc. In this case eggs can be seen, all going well, a fortnight after hatch. There are those who give a maximum time between hatching and laying but bees don't always adhere to such maxims and if the queen cell failed or the virgin got lost when on her mating flight/s by the time this is quite apparent the bees in that nuc are too old to be much further use. This then leads onto the advantage of getting mating done in a mini nuc where failures involve the loss of a minimum number of bees. The down side of this method is time, the upside possible to achieve without finding the queen.

Mini nucs

Assemble a mini nuc as per maker's instructions or by working it out from the pictures. The process described from here on refers to Apidea mini nucs but the principles would be common for the alternatives. Into each frame insert a small strip of foundation, approximately 20mm or ¾" deep, sticking it to the top bar with molten wax or PVA wood glue. The feed compartment needs to be supplied with solid food in the form of fondant.

To stock a mini nuc there are options.

Firstly the bees to stock a mini nuc need to be workers capable drawing comb, they do not need to nurture the queen cell which is already fully formed and provisioned so it is quite legitimate to take these bees from supers as there, there are only workers and no drones.

1) Frames from a super of a colony that can spare them are held over a basin and sprayed with a hand held water spray before being shaken. Some bees fly off, reckon these are the older ones, the rest the younger, fall into the basin in a wet heap. Shake in plenty of bees to ensure there are enough. The measuring mug is filled by scooping up the desired volume of bees which are then put into the mini nuc. Advantages, no special extra equipment, down side more bees than necessary are wetted.

2) Another technique is to mark a cup or mug for the required volume of bees. 250ml or 9 fluid ounces for an Apidea. Then place this mug under a large funnel such as can be bought from farm supply shops, hold a super frame by the side bar with one hand over the funnel, spray with water from a hand held small sprayer and give the frame a bump. This is most easily done by knocking the hand holding the frame with your other hand. The shaken bees fall into the funnel and into the measuring mug. Once there are sufficient bees in the mug open the floor of the mini nuc and pitch them into it. These bees will be somewhat bedraggled but soon dry out and then seem none the worse for their soaking. This funnel is either held by a helper or held in a purpose built frame which can be quite rough and ready.

Stand to hold a large funnel and measuring mug for stocking a mini nuc.

There is a school of thought that says bees to stock a mini nuc should be young workers taken from the brood chamber. In this case the queen must be found and isolated and drones need to be avoided or sieved out with a queen excluder. A more complex operation than using bees from supers which in practice have proved well able to do what is wanted of them. (Ron Brown Managing Mininucs)

If you have hatched virgin queens they can be put into a mini nuc along with the wet workers. It is more normal to place a sealed queen into a mini nuc soon after it has been stocked with bees. In the case of an Apidea it is essential that the two central frames are inserted so that the scallops in the top bar face each other to afford the space for the queen cell to hang. The mini nucs then need to be put in a cool dark place for two days for the bees to realise they are a unit and in the ideal for the queen to hatch. During this time the bees forget their parent hive and become a separate unit to locate to the place of release. When there a number of mini nucs at this stage, on occasion one queen may start to pipe which elicits responses from the other queens. This is a wonderful sound to hear. If it is warm or the bees show distress water should be sprayed on the ventilation screen. The nucs are then placed out in the open in the evening and the entrances opened making sure that the slide is fully up so as to cover the ventilation screen and prevent returning bees sitting on the screen rather than going in the entrance. If placed where exposed to full sun the little hives get too hot and the bees abscond so they should be located in dappled shade, even under a bush.

It is wise to have these mini nucs some distance away from full colonies to lessen any chance of robbing. Distance from other colonies for mating is not really a problem as both drones and queens fly good distances to the drone congregation areas, often some miles.

Once the bees in these mini nucs have been released and settled it is permissible to open the lid and observe progress through the perspex crown board. After some days if you see that comb is being drawn then there is a queen present and if not it was a dud queen cell. After a fortnight or a little longer all going well there will be eggs present sometimes more than one per cell, this is normal as the queen is still learning how. The young queen will be slim and fast moving and even though this is a tiny colony not easy to find. It is permissible to clip and mark at this stage once the queen has mated. When trying to do this the queen might escape and take flight, DON'T PANIC, just move slowly reassemble the hive and retreat while resting assured that she has been out before and knows her way home. To my mind these mini hives are too small to allow any meaningful assessment of a queen's ability so once laying these new queens should be moved on. The little frames can soon be filled with brood or stores when a flow is on, and conversely in times of dearth more feed can be needed and at this stage syrup is fine.

5. INTRODUCING A YOUNG MATED QUEEN INTO A COLONY.

Introducing a young queen straight into a full colony is risky as a full colony expects a queen in full lay rather than a beginner so in those circumstances the young queen is most often superceded. It is good practice to transfer these young queens into 5 frame nucs as a start off. The same applies to a bought in queen who needs to come to after the trauma of going through the post.

Having made up a 4 or 5 frame queenless nuc as described before there are two ways of introducing the young queen. Traditionally the queen is caught and placed in one of the many types of introduction cage. It is imperative that there is an area of such a cage where the queen can retreat so that her legs are not exposed. The worker bees outside the cage initially see this fresh incomer as a threat and attempt to attack and repel her, given the chance they chew up the bottom of her feet and render her useless as a queen's legs are used for measuring cells prior to laying and the base of her legs also secrete necessary pheromones. Once the queen has been able to solicit food and acquired colony odour she is no longer a threat and by this time hopefully the bees will have released and accepted her. The bees if left in an Apidea after removal of the queen can resort to rearing a queen from the larvae present. Some of these queens regardless of what the books say can on occasion prove to be good ones. Not to be recommended though as many such queens are poor. Ref. Ron Brown. Managing Mininucs.

A more passive introduction can be achieved in the following manner. Make up a fresh crown board for the nuc and on this place an empty Apidea with the floor opened and roof off. Mark along the front of the Apidea, down the sides and along the end of the floor as well as the perimeter of the inside of the hive. Cut out the marked area below the position of the frames and cover with newspaper. It is now possible to take an Apidea with its laying queen and place it in position on this crown board opening up the floor to the mark leaving the bees to unite through the newspaper. This avoids any stress to the queen as she remains amongst her own. Once united the bees from the nuc below move into the Apidea so it becomes jam packed. The queen does not move down onto the full frames below until a comb has been drawn far enough for her move down easily. To speed the process up give the bees a day or two then shake all the bees off the mini frames. If the Apidea is left nurse bees will come up to tend the brood and now if needed the floor can be closed off, the Apidea removed and these bees will hatch and look after another queen cell.

Uniting crown board for Apidea mini nucs. On the left mark out, centre the portion below the frames cut out and on the right ready to go.

6. MOVING INTO A FULL COLONY AND REQUEENING WITHOUT FINDING THE OLD QUEEN.

The young queen can be left in her full nuc and be permitted to build it up to full colony strength or,

The young queen having been in a 4 or5 frame nuc for a week or more will have got into full lay and so be acceptable to a full colony so can be transferred to a dequeened colony by either using a queen cage or uniting through newspaper. There is another way which is especially useful when requeening a vicious colony or when the replaced queen is surplus to requirements and time is short.

Bring the replacement queen in her 4 or 5 frame nuc and place it beside the stock to be requeened. Remove the supers from the stock and put to one side, move the brood chamber and floor of this stock off its stand and in its place put down a fresh floor with a good ridged queen excluder on top, now place an empty super or brood box on this excluder and a full sized empty brood box on top of that. Take the frames en mass out of the nuc and place in that top brood box ensuring that the queen and her cohort stay together and move all the nuc frames to one side of the brood box. Now take the frames from the surplus queen's brood box one by one and shake all the adhering bees off onto the ground in front of the set up on the original stand. Have no concern for that surplus queen and place the bee free brood frames in the upper box beside the frames and bees from the nuc. There is the need to cull as many frames as came with the nuc for reasons of space, once done put on another queen excluder and replace the supers. The full colony's bees are disorientated by such treatment. They have to squeeze in through the excluder and go up through a void to cover exposed brood while the replacement queen is protected by her own. The bees coming down from the supers are not a problem. The old surplus queen if she makes it back to the floor can't get through the excluder provided no gaps or sags exist for her to squeeze through and the empty box just above is too deep for any beard of bees below the frames to reach through the excluder so she is isolated, has no influence and becomes an intruder to be dealt with by the guard bees. There can be a dramatic overnight change in temperament in many instances. The empty box and bottom queen excluder can be removed at the next visit. As with everything there is a down side in that freshly hatched bees who can't fly tend to fail to get back but this is a small price to pay.

7. CROWNING GLORY

Having got into the swing of queen rearing there can be desire for a more permanent mark than the commonly used queen marking pens. These pens give a good mark in the year of application but the bees tend to rub this off so such marks need annual touching up. This can be a nuisance when you have special queens who could live for three years or more. The plastic numbered queen marking discs, once on, are there for life but their application is thought to be difficult especially in the field which need not be true. Take a block of wood of approximate dimensions of 150mm x 90mm x 40mm and then drill a hole in it to take the bottle of glue that comes with the kit or better still the replacement glue bottle which has the durability the original lacks. Beside this drill another hole to take a small pill/tablet container scrounged from a Chemist. Now around the edge of the block nail thin ply some 60mm high to make a tray, the idea is to stop loose items falling or blowing off. Push out three or so discs with the number facing upwards, the reason for more than one is that they tend to roll upside down with a puff of wind or a jolt and at least one disc needs to be in the right position when holding the queen. The pill container should to be half filled with water and two or more tooth picks, so that there is a spare, with blunted ends put in this tray along with clipping scissors if used. Pre loosen the lid of the glue bottle and all is ready.

With the queen held for marking, open the glue bottle and with a tooth pick, pick up a small dab of glue and place on the queen's thorax turn the tooth pick around, put the fresh end into the water and then onto a disc. The water's surface tension is enough to pick up the disc so now put it onto the queen's thorax. The wet glue holds more firmly than water so the disc stays put and the queen should now be put in an aired match box of similar small dark container to recover from the experience and allow the glue dry just as is necessary with any other type of queen marker. Once marked and in the match box allow some time to elapse then release the queen, there are differing ways of doing this. In my case I have the match box marked on one surface so I can place it on a frame with eggs and young brood then slide open the box knowing that the queen is under the inner part of the box with access to comb and hopefully she will walk onto the comb. Sometimes the queen stays in the box possibly searching for previous queens that have been incarcerated there to do battle, if so just shake her out and onto the comb and replace the frame with the queen back in her brood chamber.

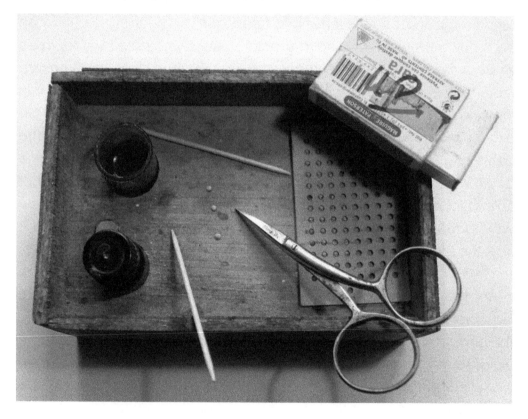

A tray for queen marking with the plastic discs. Note the fishermans fly tying scissors with man sized finger holes.

Lightning Source UK Ltd.
Milton Keynes UK
UKHW021041100522
402756UK00003B/22

9 781908 904553